HOME SPACE CREATIVE DESIGN ROUNDUPS +

居家空间创意集

浪漫雅致

ROMANTIC REFINED

深圳市海阅通文化传播有限公司 主编

中国建筑工业出版社

序 PREFACE

Just as its name implies, Romance and Elegance means rich in poetry, full of imagination and beauty and conforming to no conventional pattern, which can be perfectly demonstrated by the works meticulously included in this book. Romance is a beautiful word, inevitably arousing people's longing for beauty. People build their own ideal world and fantasy kingdom under the uncontrollable imagination. In terms of residence, everyone has his own taste and view. With the development of material culture and the blending of world culture, an ever-growing number of people begin to advocate romanticism.

Romance is not only an attitude toward life or one taste of life, but also the spice of life. Therefore, all need romantic life no matter that you are adolescent boys and girls, young couples head over heels in love, or old couples late in their life. And this book will show you magnificent romantic space one after another, bringing you a romantic experience out of ordinary. If you know romance, then I am sure that you will know the preciousness of this book. If not, then just let this book take you to enjoy a visual feast.

浪漫雅致，顾名思义就是富有诗意，充满幻想，高雅别致，美观而不落俗套之意，而本书精心收录的这些作品就是最好的诠释！浪漫，是个美丽的字眼，它无可避免地引起人们对美好的向往，不可驾驭的想象力使得他们构筑自己的理想世界，构造自己的梦幻王国。对于居所，每个人都有着自己独到的喜好和见解！随着物质文化的发展和世界文化的交融，越来越多的人开始崇尚浪漫主义。

浪漫，它不仅是一种生活态度，一种生活品格，更是一种生活调味剂，所以不管你是懵懂少男少女，还是热恋生活中的情侣夫妻，抑或迟暮晚年的老夫妻，都需要浪漫的生活，而本书将给大家展示一个个辉煌的浪漫空间，带给大家与众不同的浪漫体验！如果你懂浪漫，那么我相信你一定会懂得这本书的珍贵！如果你不懂浪漫，那么就让这本书带你享受一场视觉盛宴吧！

居家空间创意集

HOME SPACE CREATIVE DESIGN ROUNDUPS

浪漫雅致ROMANTIC REFINED

目录 CONTENTS

MONTECITO RESIDENCE, CALIFORNIA
加利福尼亚蒙太席多宅院

Architecture: Barton Myers Associates
建筑：蒙特西托小丘上的住宅

Designer： Rios Clementi Hale Studios
设计师： 里奥斯・克莱门提・黑尔工作室

Area： 313 m²
面积： 313 平方米

Interior Design/Landscape Architecture:

• Comprehensive interior work—cabinetry, kitchen/bath, wall finishes, furniture, accessories
• Program: great room, den, kitchen, master suite, guest suite, powder room, indoor/outdoor garden room
• Goal to interpret and bridge architect's intentions with client's desire for elegant home
• Honor purity of industrial architectural design
• Provide luxury, comfort, sophistication, personality through interiors
• Blur lines between interior and exterior
• Creative hybrid collection of styles and finishes to represent the clients' varied interested and tastes
• In tradition of great modern houses holding eclectic collections (see: Eames House)
• Materials, color, finishes define each space
• To support the modern open plan, all furnishings are seen in three dimensions
• Each room focuses on a different kind of experience:
  - Great Room: open, light, indoor/outdoor
  - Garden Room: textured materials
  - Den: warm, fiery materials
  - Bedrooms: peaceful, calm, soft
• Landscape to connect the architectural geometries of the house with the natural environment

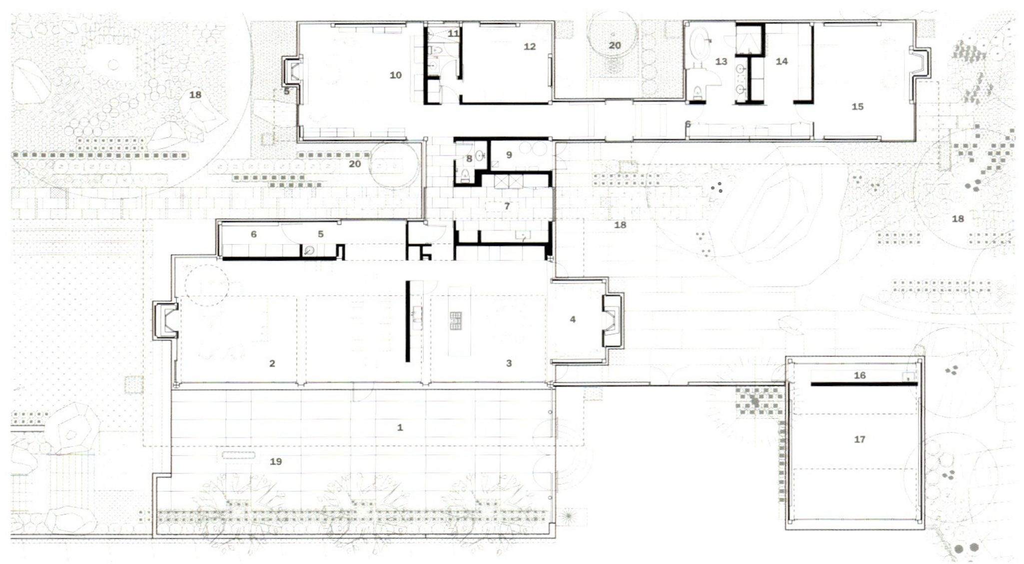

1 Terrace
2 Living/dining
3 Kitchen
4 Reading
5 Bar
6 Storage
7 Pantry
8 Powder room
9 Mechanical
10 Library
11 Guest bath
12 Guest bedroom
13 Master bath
14 Closet
15 Master bedroom
16 Garden Room
17 Garage
18 Garden
19 Firepit
20 Fountain

Plan 平面图

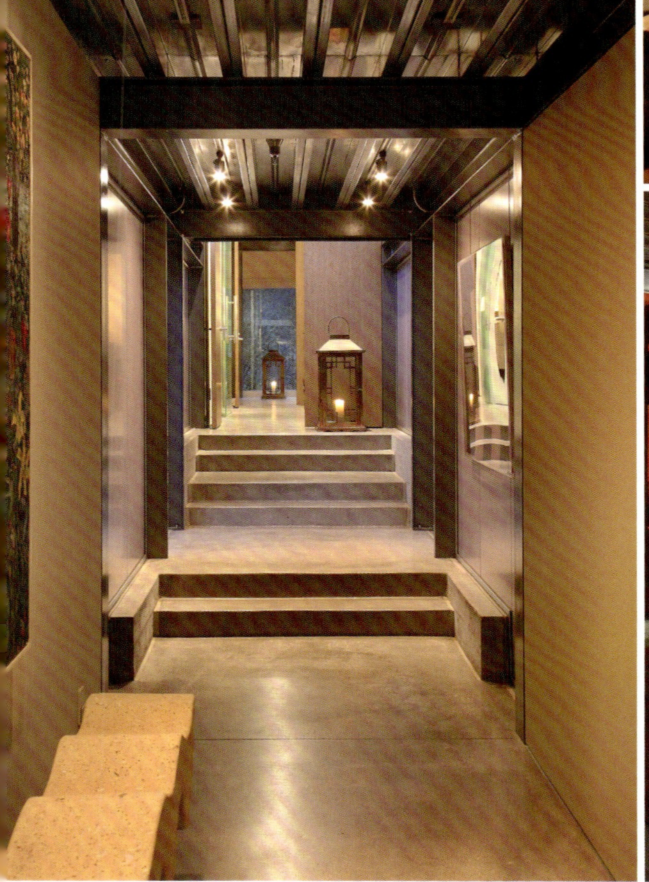

室内设计 / 园林建筑

- 综合室内工程——细工橱柜，厨房 / 浴室，墙面装饰，家具，配饰
- 项目：大间，小间，厨房，主人套房，客人套房，化妆室，户内 / 户外花室
- 目的是打造既满足顾客愿景又符合建筑师风的雅致的家
- 工业建筑设计的至上荣誉
- 通过室内设计来提供奢华、舒适、精致与个性
- 室内外设计的融合
- 创造出混合的风格，来满足顾客不同的兴趣与品位
- 拥有折中主义特色的现代与传统的结合
- 材料，颜色，成品点缀每块空间
- 为支持现代开放计划，所有的家私都以 3D 形式呈现
- 每个房间都注重各自不同的体验
  - 大间：开阔，浅色，室内 / 室外
  - 花房：纹理材料
  - 小间：温馨，炽热的材料
  - 卧室：宁静，沉寂，柔和
- 将自然环境与房子的建筑几何图形有机结合

SECOND-STAGE MODEL HOUSES OF FUTONG TIANYI BAY — CLASSICAL NEW SPIRIT

富通天邑湾二期样板房——古典的新精神

Design company：Shenzhen Haoze Space Design Co.Ltd.
设计公司：深圳市昊泽空间设计有限公司

Designer：Han song
设计师： 韩松

Location：Dongguan, Guangdong
项目地点： 广东东莞

Classical New Spirit—Building 3, House Model A

Arrangement of interior space is beautiful and romantic. The collocation of white and cream-color creates a sense of pure and noble beauty. Moreover, exquisiteness and elegance are fully interpret through the details including traditional fireplaces, fashionable furnitures, mellow lighting and crystal material. There is not only the continuation of history but also logical thinking not rigidly adhering to traditions, light and bright tones as well as clear texture, creating distinctive European romance.

古典的新精神——3 栋 A 户型

室内空间的布置唯美浪漫，用白色和浅米色搭配，营造出纯净高贵的美感，并通过传统壁炉、时尚家具、柔美灯光、水晶材质等细节，演绎精致高雅，既有对历史的延续又不拘泥于传统的思维逻辑，浅淡明快的色调，清晰的质感与机理，营造了不一样的欧式浪漫。

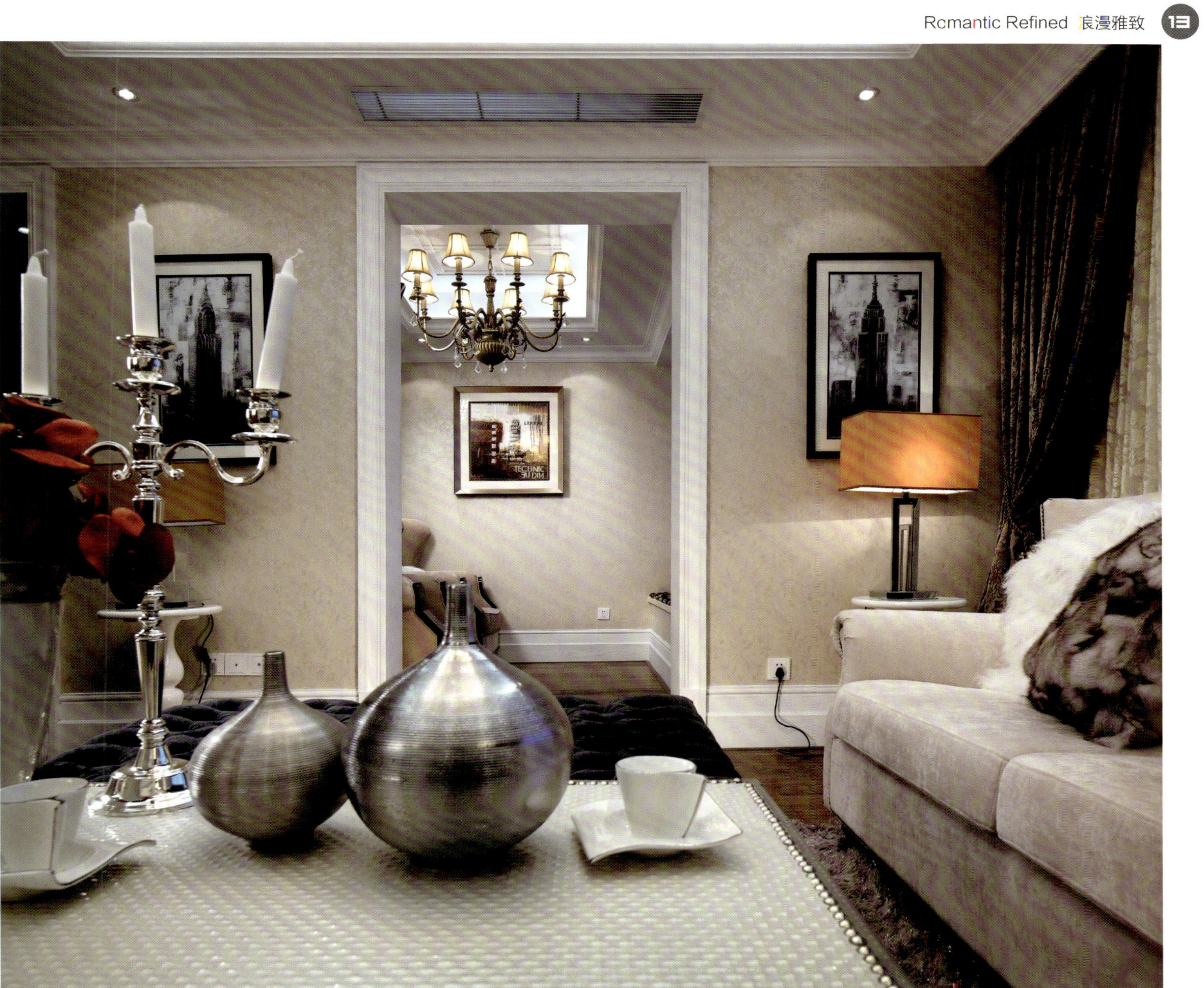

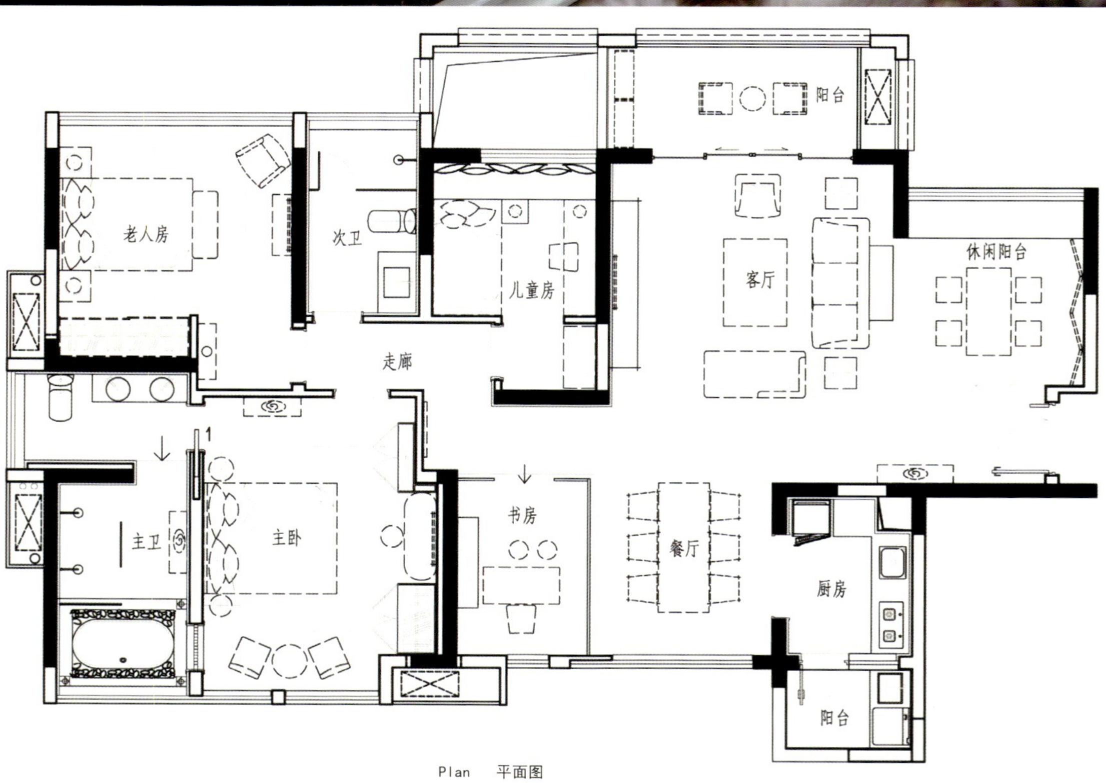

Plan　平面图

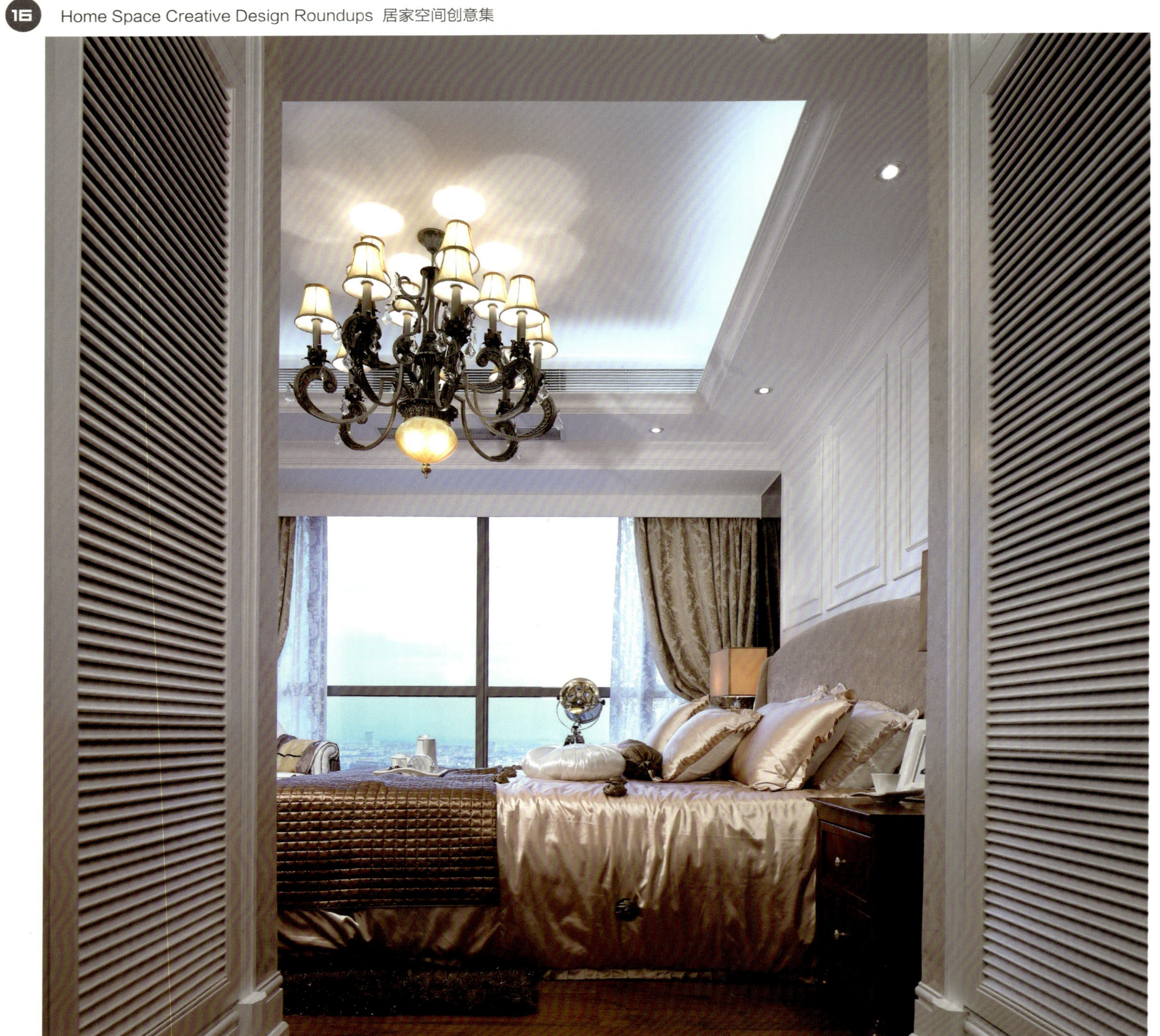

POLY (CHONGQING) GOLF GARDEN THREE PERIOD
保利（重庆）高尔夫花园三期

Design company：Dian Kuo space design company
设计公司： 点廓空间设计

Location：Chongqing
项目地点： 重庆

Area：300 m²
面积： 300 平方米

Neoclassical French style simplified classical palace design,it is classical,original and inherits classics.It reserves the essence of classical architecture and has three features.firstly, symmetrical layout creats magnificence and luxury living space;Secondly,noble style reflects elegant and dignified.Last are details,silver mirror,white caving totems,crystal containers are all exquisite.White wallboard wireframes match fresh light blue,elegant purple,pleasant grayish green perfectly in this style,creating a romantic Spring of Paris.

法式新古典风格即简化了的传统法国宫廷式设计，古典、原味，传承经典。它在简化那些繁复设计细节的同时，又保持了传统法式建筑的精髓，主要具有三个特点：一是布局上突出轴线的对称，恢宏的气势以及豪华舒适的居住空间；二是效法贵族风格，高贵典雅；三是陈设细节处理上运用银镜、白色雕花图腾、水晶器皿等，制作工艺精细考究。本案的浪漫法式新古典，运用了白色墙板线框造型结合清新的浅蓝色、优雅的浅紫色、怡人的灰绿色作为点睛色，营造浪漫巴黎春天风情空间。

ROXBURY RENOVATION

罗克斯伯里翻新的设计

Designer：Daniel Torres，Melissa Bacoka
设计师：丹尼尔·托雷斯，梅丽莎·巴克卡

Location："The Flats" of Beverly Hills, CA
项目地点：加利福尼亚比弗利山庄"The Flats"

Area：929 m²
面积：929 平方米

The designers looked to the existing elements of each room to solve the program concerns of the clients. From the brick exterior, with custom-designed sconces, into the entry, the home exudes a gentle manner, with each room gaining intensity as one progresses through.

Furniture purchased from antique auctions around the world include superlative pieces by Gio Ponti, Edward Wormley, Piero Fornasetti, Paul László, Charles Hollis Jones, and Karl Springer. As needed, pieces were restored and reupholstered, but many were kept intact, such as the turquoise-blue breakfast set by Fornasetti, sporting fanciful images of sea life. Rooms are no neutral backdrop for such fine pieces, as each one is heavily treated in surface and color.

Starting off quietly, the living room is a mélange of pale shades and surprising textures. From there, rooms become more saturated in color and application. Dining rooms are surrounded by leather walls, with the main room in brown leather as background to the long, high-gloss wood table, and the breakfast area in bright white to correspond with the marble table top. Both tables sport the same dining chairs, but with a reversal of leather: main chairs in white and breakfast chairs in brown.

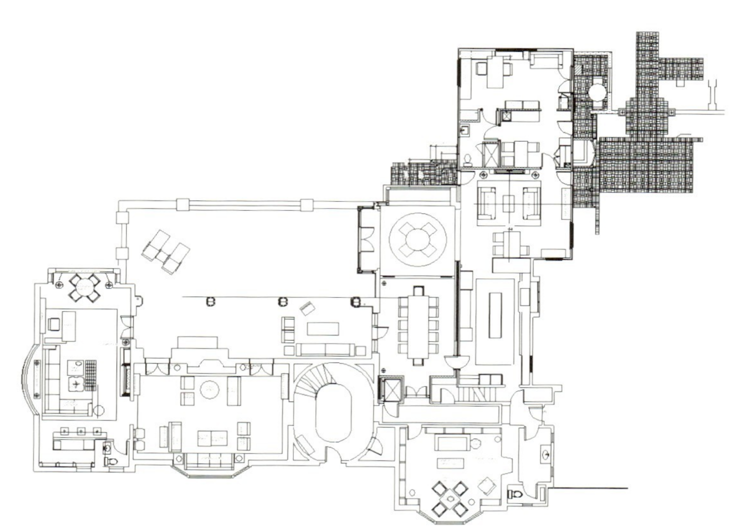

The first floor plan 一楼平面图

设计师期望用每个房间的现有元素来解决客户对项目的担忧。从外部的砖墙，沿着特别定制的壁灯的光晕，走进去，整个房间散发着一种柔和的气息。随着施工的不断进行，一间一间地慢慢呈现。

室内的家具是从世界各地古董拍卖会买来的设计大师的顶级杰作，如吉奥•庞帝、爱德华•奥姆利、皮耶罗•弗纳赛提、保罗•拉伊克、查尔斯•霍利斯琼斯和卡尔斯•普林格等。根据设计需要，对它们进行修复和重装，但是其中大多都会保持原貌，比如弗纳赛提的一套描绘了海洋生物的各种稀奇古怪形象的蓝绿色餐具。鉴于每一件家具的表面和颜色都较厚重，房间的背景也没有保持中性基调。

一切始于静谧，起居室是浅色调和惊人质感的结合。从这里开始，房间在色彩和应用上变得更加饱和。皮革墙面围成餐室，主室用棕色皮革作为一张长形、高光泽的木桌。早饭区用的是亮白色与大理石桌面相契合。桌子都配上相同的餐椅，但是所用皮革正好相反：主室椅子采用白色，早餐区椅子则采用棕色。

ROBISON LOFT

罗宾逊 LOFT

Designer：Jim Poteet, Brett Freeman
设计师： 吉姆•波提特，布雷特•弗里曼

Location：San Antonio, Texas
项目地点： 圣安东尼奥市，得克萨斯州

Area：362 m^2
面积： 362 平方米

This project is the interior finish out of a 3900sf raw shell space in a rehabilitated 1920s era factory building at the southern edge of downtown San Antonio. The client is an internationally popular musician who desired an urban retreat when not touring or recording.

The design groups service areas in the center of the overall plan, leaving the abundant windows –all with city views --for the major living spaces. This central zone is clad in white oak siding which lends a natural warmth to the public spaces. The kitchen, recessed into this service zone, is also open to the large living and dining space. A 30' long island with stainless steel top and sides houses the cooking and washing functions and integrates seating and storage as well.

Custom steel interior windows and doors, painted the same gunmetal as the existing serve as a unifying material throughout, as do the dark concrete floors which have been polished to a reflective finish.

Care was taken to use the massive concrete columns as elements in the design—in the entry hall, the kitchen, master bedroom and particularly the master bath, where the double shower wraps around a column.

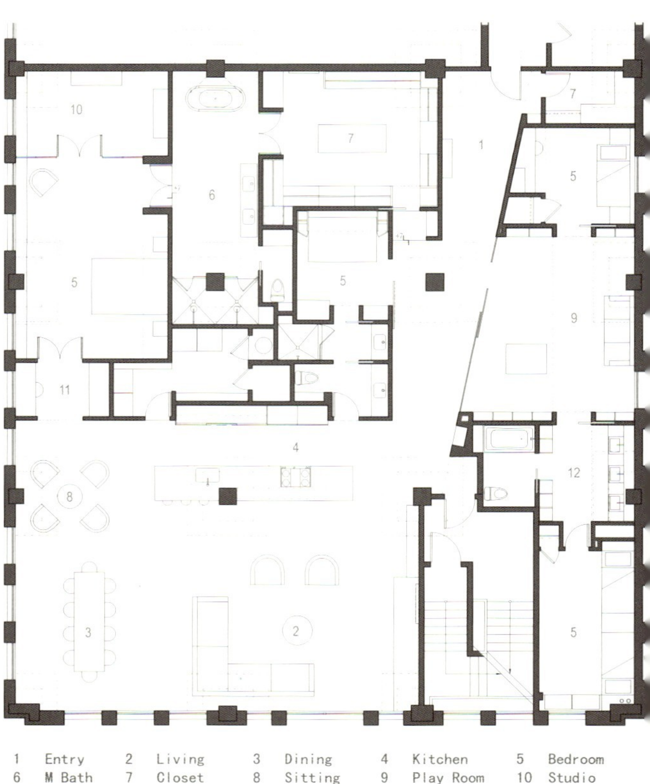

Plan 平面图

此项目是在一块 3900 平方英尺（约 362 平方米）的未开发的贝壳形状的土地上开发的，它是在圣安东尼奥市市中心南端的一座 20 世纪 20 年代的一家工厂建筑的基础上重新进行的室内装修。客户是一位享誉国际盛名的音乐家，他想在不巡演或录片之余，有个休息的寓所。

此设计集中心整体规划服务区于一身，将大量的有城市观景的窗子作为主要的生活区。中心区由白橡木覆盖，这样，为公共区域增添一丝温暖。镶嵌式的厨房，对外向起居室和餐厅开放。厨房是 30 米的长岛款式，不锈钢的顶部与四周集做饭、涮洗、坐及存储功能于一体。

定制的室内装修用的钢材窗户和门，漆成统一的青铜色，与被抛光后的暗色混凝土地板交相辉映。

在设计中使用大量的混凝土罗马柱——在门厅、厨房、主卧，尤其是在主浴室中，双人淋浴围绕着罗马柱，此设计可谓匠心独运。

SINGLE-FAMILY HOUSE MORASKO
MORASKO 的独栋别墅

Location： Poznań, Poland
项目地点： 波兰，波兹南省

Area： 300 m²
面积： 300 平方米

Arranging the interior of a large, impressive residence located in a high class residential area of the city was an interesting task. The investors, who bought a classic, manor-like (with a porch and two columns in the front) house design, unanimously decided that the interior should be – contrary to the outside of the building – modern and spacious, without redundant knick-knacks, exaggerated forms or 'heavy' furniture.

In this way, our main challenge was to unite the two opposite poles.

The then-new interior decoration fashion for bold compositions of baroque crystal chandeliers with extremely ascetic furniture shapes and materials appeared on our way, becoming an unlimited source of inspiration.Works on the interior of the house began as early as on the stage of foundation casting and wall construction – which is why we could avoid the time-consuming modifications required by ourdesign. The design and execution process took three years.

The owners decided to use quality finishing materials, leaving the designers a lot of freedom in shaping the style and atmosphere of the interior.

Actually, all pieces of furniture and more distinctive interior decoration elements have been especially designed for this particular space – starting from sliding glass doors, stairs and balustrades,through kitchen furniture, simple cupboard and fancy tables in the living room, bathroom furniture,to lacy mirrors in the hallway and kitchen hood clad with mirrors. The fact that all made-to-order elements originated from local companies and craftsmen gave us the largest possible freedom of creation and the possibility to freely manipulate and play with the style. That is how cotton-candy small tables with classical shape, made of a totally non-classical (high gloss painted mdf board) materials appeared in the hallway.

The flooring on the ground floor was made of polished cream-coloured gres tiles, while on the stairs and first floor – of merbau wood.For our general satisfaction, we also added some well known icons/pearls – Ghost chairs, Axor StarckX tap, Bisazza mosaic on the kitchen wall and …electrifying photographs of Marilyn Monroe.

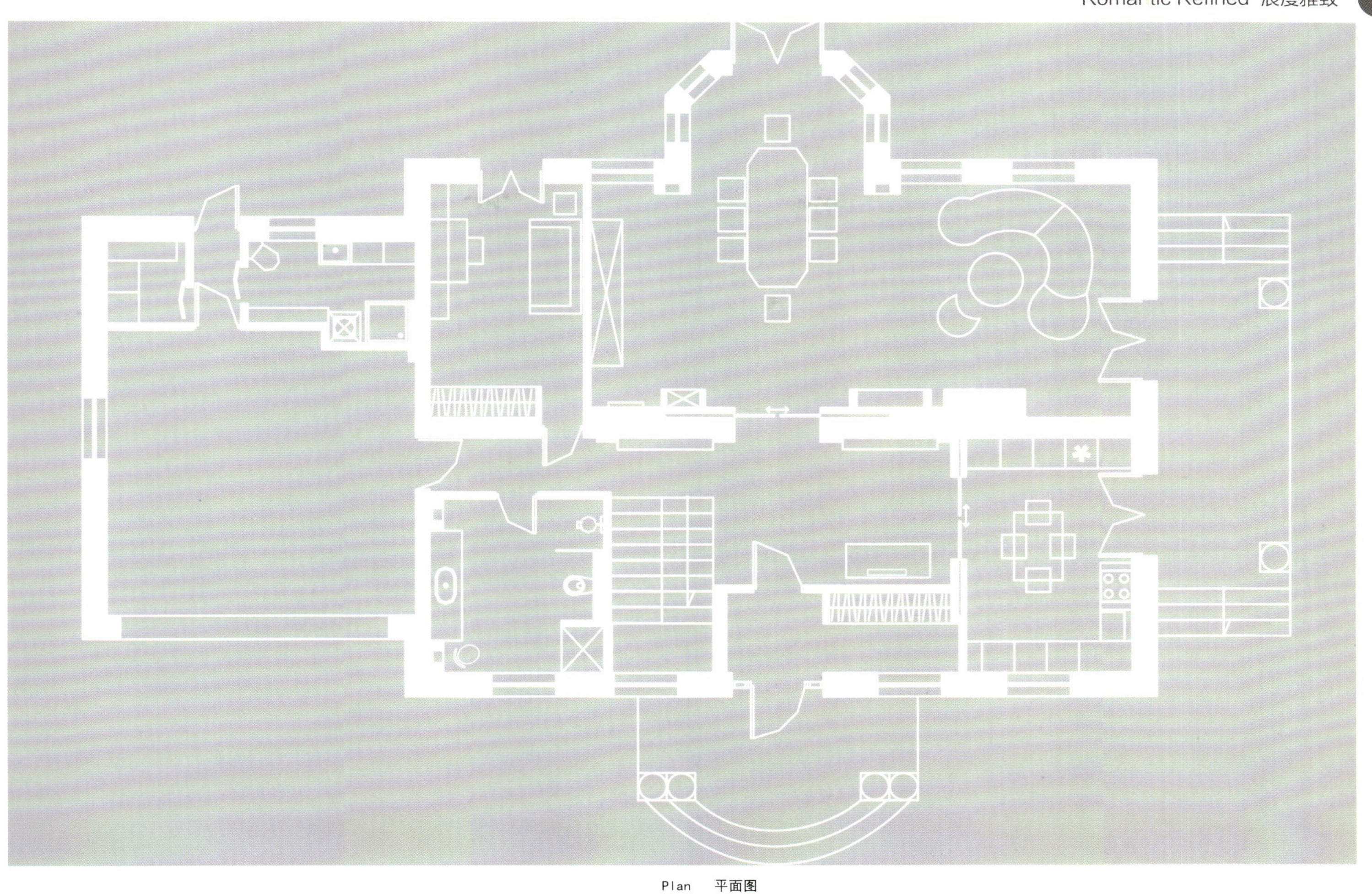

Plan　平面图

为一幢位于城市的高级住宅区的房子作室内设计是一件很有乐趣的事情。业主买下了一个古典风格的庄园（前面有走廊和两根罗马柱），设计师觉得室内设计应该与建筑外部环境形成反差，不需要过多的繁缛装饰、样式夸张或是笨重的家具，而是要体现出现代气息和开阔的空间。

这样一来，最大的挑战就是如何让这两个截然相反的极端风格和谐统一。

当时室内装饰新潮流将巴洛克水晶吊灯和极端苦行主义家具造型和材质大胆结合，设计师用自己的方式将这种潮流再现，成为灵感的无限源泉。房屋室内设计早在地基铸造和墙体建筑阶段就开始了，这也是设计师能避免掉耗时地修改设计稿的原因。设计和施工一共用了三年的时间。

主人决定使用高质量的装修材料，这使得设计师在设计风格和营造氛围时有很多自由的发挥空间。

事实上，所有的家具和特色鲜明的装饰元素都是为这个空间而特别定做的，从滑动玻璃门、楼梯、扶栏、到客厅里简约的柜子和精致的茶几、浴室，再到门厅处的花边装饰镜以及光洁闪耀的厨房，皆是如此。所有的定做的元素都出自当地公司和工匠，这充分为设计提供了自由创作和风格定位的可能性和可操作性。所以可以看到摆放在门厅处的那张用现代材料（高光中密度纤维板）做成却尽显古典风格的棉花糖样式的桌子。

一层采用亮面米色格蕾丝瓷砖，而楼梯和二层则是印茄木。为了整体满意度，也添加了一些亮眼的元素，如幽灵椅、斯达克水龙头、厨房墙面的碧莎马赛克以及玛丽莲·梦露的性感照等。

SWEET DREAMS

甜蜜梦境

Designer：Jamie Drake 设计师：杰米·德雷克

Entering a Jamie Drake-designed room, be it modern or historic, can make you feel as if you're walking into a bold and brazen hot-house or a subdued and serene spa. The most hushed of rooms—bedrooms, libraries—become lush and lively in his deft hands. Gathering rooms morph into restful havens without sacrificing a sense of surprise or sociability. The Drake touch is always evident.

The Bedroom Suite: Sweet Dreams

The ladies who lunch also need to sleep. Candy-colored and rich in design, this confection of a bedroom is sure to instill sweet dreams. High glamour, high color, and high gloss are the motifs.

From the upholstered walls to the bed cover to the upholstered chairs, the room is wrapped in Jamie Drake's fabric line, Jamie Drake for Schumacher. The fabrics for Schumacher have come together in a delicious parfait of color and texture. The walls are upholstered in an orchid-colored fresh interpretation of moiré with an embossed dot pattern. A cool base of lush lavender-tinged gray wool carpet from Patterson, Flynn & Martin tempers the room's exuberant floral palette. Curtains in Drake's signature "Jazzed Stripe" fabric from Schumacher hang from elegant crystal rods." The custom sleigh bed is both severe and romantic at the same time," says Drake. "Its square lines were explicitly envisioned to show off my "Not Square" fabric, a silk velvet cut to reveal a soft metallic ground." Gilt nailheads further emphasize the bed's form. On each side of the bed, an Alpha Workshops black wood "Waterfall" table with fuchsia peering through a negoro nuri lacquer finish holds an antique Chinese lamp from Mallett's. Above the headboard, a painting by David Mann depicts a celestial explosion of energy. "I love the palette, its modernity and its mystical feeling," says Drake. A Russian 19th-century desk in birch, black, and gilt verre eglomise and gilt bronze mounts come together in a very elegant design. In the dressing room custom smoky lavender lacquer cabinetry to house the clothing, shoes and accessories.

走进一间杰米·德雷克设计的房间，无论是充满现代感还是富于历史感的，都会使你感觉像是踏入一间温室或一处柔和宁静的温泉。房间最安静之处——卧室和书房——在他的妙手下变得生动热闹起来。他将房间变成宁静的心灵港湾而又不乏惊喜感和社交功能。德勒克风格总是如此显而易见。

卧室套房：甜蜜梦境

除了吃饭，女士们也需要睡觉。一个糖果色又富于设计感的宛如糖果般的卧房会构筑一个个甜蜜的美梦。魅力四射、色彩鲜明和光亮炫丽就是设计主旨。

从软垫墙到床罩再到软垫椅，整个房间都被包裹在杰米·德雷克独特的面料中——舒马赫面料。舒马赫面料汇集到一起，仿佛一道由色彩和质地构成的美味甜点。墙上装饰着清新的淡紫色云纹软垫，上面点缀着压花波点图纹。染成淡紫色的灰色羊毛地毯是出自帕特森、弗林 & 马丁，这种冷色调调和着这个房间丰富多彩的花卉调色板。窗帘是用舒马赫家的德雷克签名“活力条纹”的面料，从高雅的水晶棒上垂落下来。德雷克说：“定做的雪橇床庄重的同时又不失浪漫。其方形线条如预期的那样鲜明地突显了我的‘非方形’面料，就好比撕开一条丝绒来展现柔软的金属地面。”镀金的钉头进一步突显床形。在床的两边，是阿尔法工作室的黑檀木“瀑布”床头桌，用 negoro nuri 涂漆抛光打蜡，呈现紫红色。桌子上摆着玛丽特家的中国古董灯。床头板上是一幅大卫·曼恩的画，上面描绘了天体爆炸。德雷克说：“我爱它的色调，它的现代感以及它的神秘感。”以高雅的设计，桦木、黑色的夹金玻璃画屏和镀金的青铜裱框汇聚一起，造出了一张俄罗斯 19 世纪的桌子。在更衣室里，摆放着定制的烟熏淡紫色光滑表面的细木家具，用来装衣服、鞋子和配饰。

YORK HILL
约克山住宅

Design Company: lux design

设计公司： 勒克斯设计

When the clients come to LUX Design, they had in their hands 5 panels of the artwork now in the living room. They had found the pieces online and fell in love with it immediately. As a result, they wanted a house designed around the piece of art. The result was a transitional home with hints of both modern and baroque pieces.

The whole house was completely redone. Walls came out, new ones went up, and custom pieces finished off the space. The bed in the master bedroom was built just for them. The bed frame was custom built including the tufted headboard. The completely new bathroom included a custom vanity that was built higher than the standard 36" to make it more comfortable for the tall family. The pipes that could not be moved were covered with cabinets, which doubled as extra storage.

As part of the renovation, the kitchen was also completely ripped out, and in the process the kitchen and the dining room were combined for an open concept feel. It made the dining room feel more spacious and luxurious.

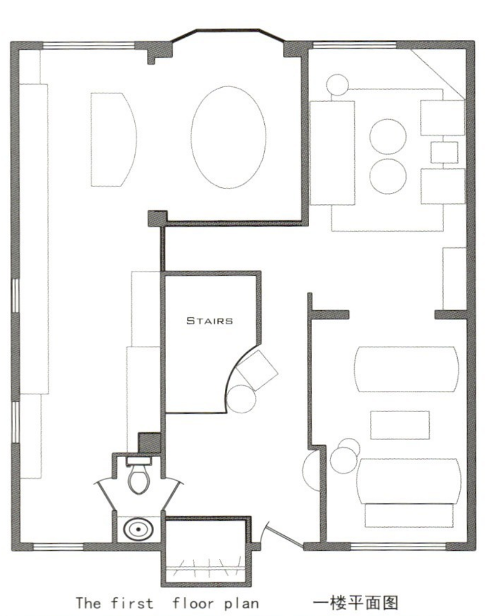

The first floor plan 一楼平面图

To continue the feeling of spaciousness, the underside of the stairs were painted white to blend into the wallpaper. This visually rose the stair 1' and gave the entrance a touch of subtle luxury.

当顾客来到勒克斯设计公司的时候，他们手中会拿到陈列在客厅的 5 份艺术品的面板。他们在网上搜到这些艺术品，并深深地爱上它们。自然而然地，他们想要一套包围在艺术品中的房子。他们的房子就显露出现代和巴洛克的艺术作品。

整座房子被彻底地改头换面，新砌的墙、定制的元素出现在新的空间。主卧室的床是特别设计的。床架，包括植绒的床头板都是专门定制的。这个全新的浴室包括一个定制的化妆台，是建立高于标准 0.91 米，目的就是为高个子的家庭提供舒适的空间。那些不能被拆除的管道都用小柜子盖住，这样也能将存储空间扩大一倍。

作为翻新设计的一部分，厨房也焕然一新。厨房和餐厅的完美结合，给人以开阔感，同时提升了餐厅的开阔与奢华。

为延续开阔的空间感，楼梯台阶下面被漆成白色，与墙纸融为一体。这样，从视觉上提升了台阶的高度，并让入口处有一种简约且奢华的感觉。

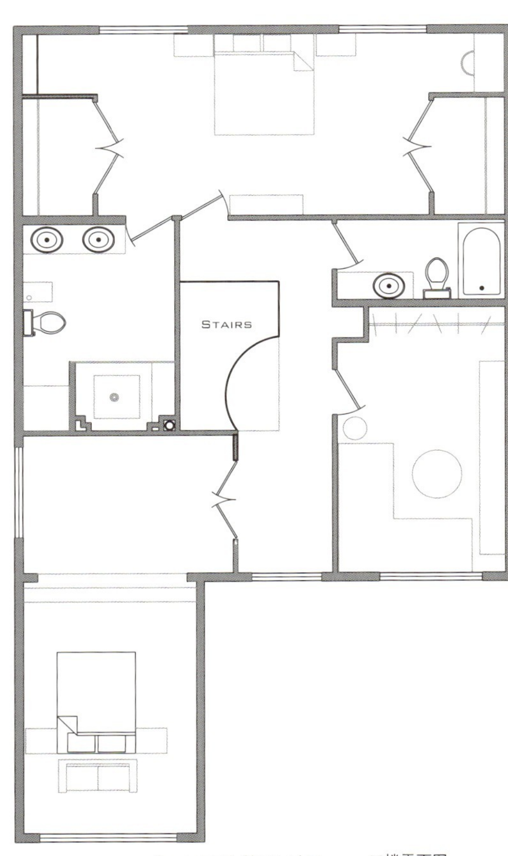

The second floor plan　　二楼平面图

AUSTRIAN-STYLE VILLA

奥地利风格别墅

Designer: Zheng Zhihao
设计师：郑志浩

Location: Taiwan
项目地点：台湾

Area: 396 m^2
面积：396 平方米

This Austrian style villa has been famous scene in film and television circle since it has been finished. Every year, many films are shot here. In the living room, the wall is covered by warm yellow Spanish wallpaper, and the floor is polished and tiled in double colors, so as to make it spacious. In front of the living room, there is a pink flower house connected with the garden, paintings by Mu Xia, Italian mirror and hand painted furniture are placed there. There is a porch at the connection point between the back of the living room and the backyard. Small as it is, it has a function of preventing the daylight.

The hallway is very necessary. At the right, there is a washroom, shoes room and changing room. On the left, there is a antechamber, with a marble fireplace. All of the things mentioned belong to the new classical architecture, while the dining room belongs to the new artistic style, with a lot of rosewoods and fold foil boxes. All of the wall lamps and droplights are newly bought classical antique lamps.

The room on the second floor is for painting. It combines both the Chinese and western style, especially the Italian style while designing for the sake of Chinese paintings and the storage of the Chinese paintings.

Living space, including bathroom, video room, master bedroom, powder room and changing room is on the third floor. The hallway is in a classical style, decorated with the roman columns and modeling mirror. The living and video room are in American style, after all, the living conditions in America is the most comfortable in the world.

这间奥地利式的别墅，完工后几乎在影剧圈中成了知名场景，每年都有许多影片在此拍摄。客厅采用暖黄色西班牙壁纸做墙面，地面采用双色抛光拼贴以造出空间感，在客厅前方与前花园相接处设有温馨的粉红色花房，放置慕夏的画作与意大利的镜子及手绘家具，在客厅后方与后院相接处设有后廊，虽然不大但是对于大面的采光还是有隔离的作用。客厅与厨房可以相通，这使得生活十分便利。

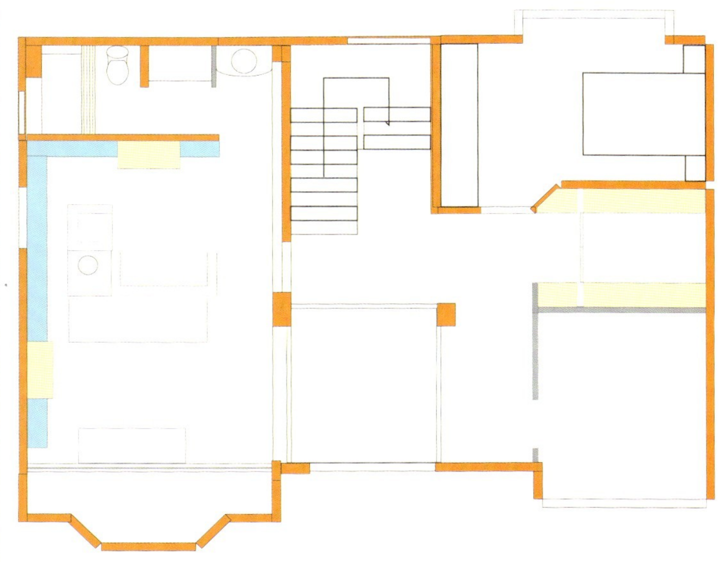

The second floor plan　　二楼平面图

门厅是绝不可少的，门厅右侧有卫生间、鞋室与更衣室，门厅的左侧则是前厅，有大理石壁炉作为端景。以上都是属于新古典式，餐厅则是新艺术风格，大量采用花梨木与金箔筐，就连壁灯与吊灯也都是采购新古典式样的古董灯放置的。

这一间的二楼是作为画室使用的，因为业主画国画并收藏了许多国画，所以在设计上采用中国式与意大利式混搭，中西风格融合在一起。

三楼是主要的生活空间，以浴室、视听室及主卧室、化妆间、更衣室为主，三楼门厅采用新古典式样，搭配罗马柱与造型镜，卧室与视听室则是采用美式风格，毕竟美国居家在卧室设计的舒适上是全球最佳的。

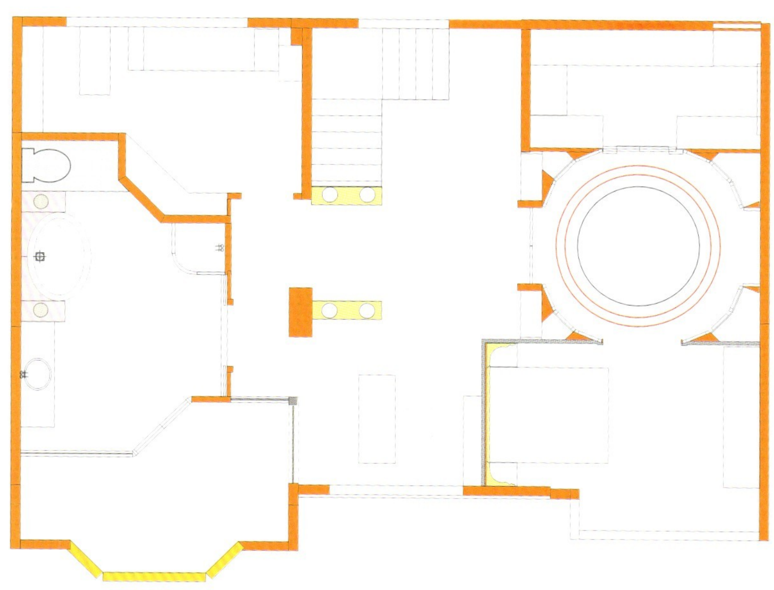

The third floor plan　三楼平面图

LINKOU GERMAN TYPE VILLA
林口德式别墅

Designer：Zheng Hongze
设计师：郑闳泽

Location：Taipei
项目地点：台北

Area：350 m^2
面积：350 平方米

The owner is a young entrepreneur,he often visites to European exhibition and prefers to Germany style.He intends to build a villa in his own land and live with merontic mother.They form a European habit in life and diet,so it must be sophisticated both in form and connotation.The first floor forms a European style which is controlled by host.Family party area,breakfast bar,reacreation zone,Christmas decorations and remote switch of different situations are all for kids.

The second floor mainly consists of master room and is controlled by hostess.Because of the limitation of height,we use modern style instead of European.The couple are interested in dressing,so both of them have a private dressing room,bath room is very spacious ,either. The third floor consist of children's reacration room,computer room and storage room.

Baroque living room is furnished by DranaWhite marble columns.Wall panels are all custom,floor use seamless Tino marble,cabinets are special custom of American style.Three sides of dining room are coverd with glasses inlaid copper. All of the woodenworks are painted with chitin. The bedboard of the master room is a custom carving which is valued 190,000 yuan.Bath room installs an auto-door.The first floor is Baroque style and the second floor is neoclassical style which is low-key luxury.

业主为年青企业家，经常去欧洲参展，偏好德国样式，由于母亲年迈所以自建一处别墅，并将母亲接来同住。由于生活习性与饮食习惯均十分欧化，所以在设计上从形式到内涵均十分讲究，一楼属于男主人决定的空间，式样偏德奥的欧风样式，由于子女尚小，所以预留许多家庭派对、早餐吧与节庆娱乐空间，还有圣诞灯挂饰设计以及情境遥控开关。二楼以主卧室为主，属于女主人决定的空间，同时由于高度限制难以展现欧洲风格，所以设计上较为现代些。男女主人都喜欢穿着打扮，所以各有一间专属的更衣室，浴室也十分宽敞舒适。三楼则是子女的活动空间与计算机室、储藏室。

巴洛克式的客厅搭配银狐大理石柱，壁面饰板均为开模定制品，地面采用帝诺大理石做无缝处理，橱柜采用特殊开模订制美式橱柜，餐厅三面覆层镶铜玻璃，所有木作均做甲壳素涂布处理。主卧床头 19 万元雕刻定制品，浴室采用自动门装置。设计样式，一楼巴洛克，二楼新古典低调奢华风。

THE MONACO CONVENTION

摩纳哥之约

Design Agency：Xingyutian Design Company
设计公司： 行于天设计公司——石子出品高端工作室

Designer: Shi Xiaowei Kong Weiduo
设计师： 石小伟 孔魏躲

Location：Lvcheng, Nantong , Jiangsu province
项目地点： 江苏南通市绿城

Light cream-color and aesthetic furnitures bring romance to the whole space. Harmonious color and design can be seen everywhere. Good features are added by the decoration to the point and the lighting. And romance is created by high-quality family property, soft lighting and crystal materials. The collocation of color and classic and romantic furniture are the most important part of the design.

浅米色调且形状唯美的家具让整个空间充满浪漫的气息，每个地方都讲究颜色和谐呼应，所以整个设计看起来很和谐，恰到好处的装饰和灯饰照明给整个设计增色不少，高质感的家私、柔美灯光、水晶材质等细节营造出足够的浪漫。色调的搭配和古典浪漫的家具是设计中的重点。

MODEL HOUSES IN XUHUIYUFU 9A, CHANGSHA
长沙旭辉御府 9A 样板房

Designer：Peng Lin　Chen Zexin
设计师：彭林　陈泽鑫

Finishing time: July, 2012
竣工时间: 2012 年 07 月

Area：103 m^2
面积：103 平方米

The major tone of this design is fresh Postmodern European style, combined with classics and fashion, grace and personalized. Retro lines with fashionable color and quality, showing a sense of elegance and grace of modern well-bred female white collars. The designer combines the tradition with the modern beauty, letting out another kind of grace and aesthetics, by using his artistic deposits, mind of innovation and valued attitude.

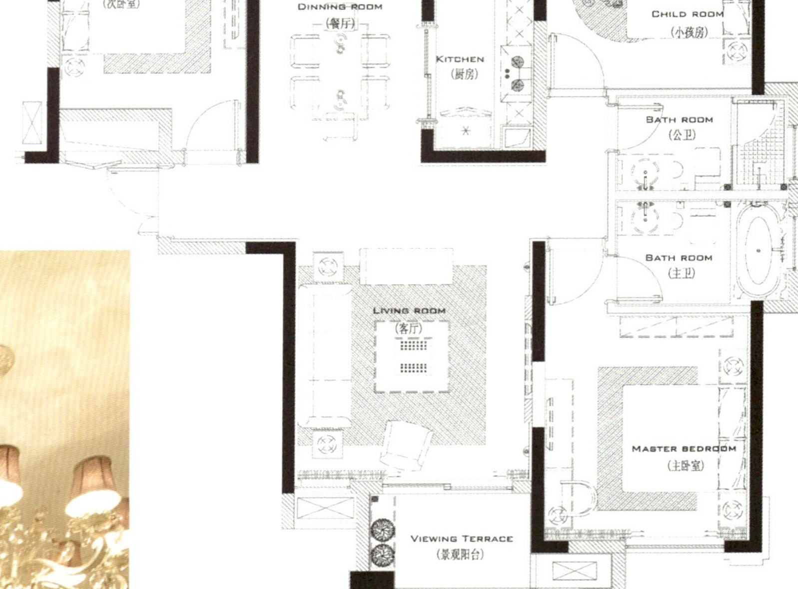

Plan　平面图

本案设计以清新的后现代简约欧式为主调，经典中透出时尚，优雅中显露个性，复古的线条样式，点缀着时尚的色彩与质感，犹如都市知性女白领那样清新典雅、高贵大方。设计师以丰富的艺术底蕴，开放创新的设计思维以及尊贵的姿态，奢华不俗又独特不羁，让昨日的经典传承今天的艳魅，散发出另一种雅致与唯美。

Master Bedroom B Elevation Drawing　主卧室B立面图

图书在版编目（CIP）数据

浪漫雅致 / 深圳市海阅通文化传播有限公司主编.
北京：中国建筑工业出版社，2013.4
（居家空间创意集）
ISBN 978-7-112-15186-8

Ⅰ.①浪… Ⅱ.①深… Ⅲ.①住宅—室内装饰设计—图集 Ⅳ.①TU241

中国版本图书馆CIP数据核字(2013)第038827号

责任编辑：费海玲　张幼平　王雁宾
责任校对：姜小莲　陈晶晶
装帧设计：陈秋娣
采　　编：李箫悦　罗　芳

居家空间创意集
浪漫雅致
深圳市海阅通文化传播有限公司　主编
*
中国建筑工业出版社出版、发行（北京西郊百万庄）
各地新华书店、建筑书店经销
深圳市海阅通文化传播有限公司制版
北京方嘉彩色印刷有限责任公司
*
开本：880×1230毫米　1/16　印张：$5^1/_2$　字数：180千字
2013年 5月第一版　2013年 5月第一次印刷
定价：29.00元
ISBN 978-7-112-15186-8
(23277)